101 *one hundred one* SIGNS *of* DESIGN

TIMELESS TRUTHS FROM NATURE

one hundred one 101 SIGNS *of* DESIGN

TIMELESS TRUTHS FROM NATURE
JOHN D. MORRIS

First printing: August 2002

Copyright © 2002 by Master Books, Inc. All rights reserved. No part of this book may be used or reproduced in any manner whatsoever without written permission of the publisher, except in the case of brief quotations in articles and reviews. For information write: Master Books, Inc., P.O. Box 726, Green Forest, AR 72638.

ISBN: 0-89051-367-8
Library of Congress Catalog Card Number: 2002105383

Printed in the United States of America.
Please visit our website for other great titles:
www.masterbooks.net
For information regarding author interviews, please
contact the publicity department at (870) 438-5288.

INTRODUCTION

"The heavens declare the glory of God" (Ps. 19:1), and so does the earth, even plants and animals. Certainly the greatest use of the tongue and the printed page is to do so also.

This little book contains 101 brief quotations from my various writings which speak of this incredible design in nature. Scripture teaches that by studying "the things that are made," the nature of its Creator, "even His eternal power and Godhead" are "clearly seen." They are so clearly seen in fact, that if one fails to recognize and acknowledge the Creator's hand behind it, they are "without excuse" (Rom. 1:20).

Rightly understood, the intricacies of any form of life testifies to the creative handiwork of God. Furthermore, the nature of earth's mineral make-up and landforms testify both to its creation and the great flood of Noah's day which restructured it.

Today we have the treasured privilege of observing the Creator's creative majesty, knowing Him more fully through it, directing others to Him, and giving Him the glory He deserves. To these ends, this book was compiled.

You might not have heard it in the media, but discovery after discovery confirms **the truth of God's Word** *and the benefits of living according to His guidelines. The problem for Christianity is gaining access to this revealing information, for many educators, politicians, and media outlets have joined forces to continue promoting the evolutionary, humanistic, naturalistic world view.*

2

There are two categories of evidence which I keep coming back to as confirmation of creation: (1) that of the *incredible design and order* in living systems, and (2) the *separateness of the basic body styles* of plants and animals in the fossil record.

3

Regarding design, even the smallest single-celled organism, the kind which evolutionists say are similar to those which evolved spontaneously from non-living chemicals, is complex beyond our own ability to understand, let alone recreate.

4

While the study of living systems gives us evidence in the present that life could not have arisen by chance, a study of the fossil record gives us an indication as to what that life was like in the past.

5

The fossil record shows (1) that life forms manifest little or no change during their history, (2) that most fossil types are virtually identical to their living descendants, (3) that fossil types appear in the fossil record without ancestral lineages, and (4) that fossil organisms either became extinct or have survived into the present. This is exactly what should be present if creation has occurred.

6

The fossil record, while vaguely compatible with evolutionary stories, clearly favors the creationist interpretation.

7

The real key for resolving the creation/evolution controversy is in a study of the age of the earth. Evolution demands long periods of time, but if the earth is much younger, **as the Bible teaches,** *then evolution is even more foolish.*

8

I am convinced that the surface of the earth appears to have been shaped and remolded in the past by an incredibly dynamic watery cataclysm.

9

If the Flood happened, then it would have accomplished great geologic work. It would have eroded material from one location and deposited it in another, and in those muddy sediments would be animals and plants which had died in that flood.

one hundred one SIGNS of DESIGN

10

If the biblical flood really happened as the Bible describes, then we would expect the geologic results of that flood to show their cataclysmic origin, the result of processes operating on rates and intensities far beyond those operating in the present, and they would be operating on a regional scale.

As creationists have been maintaining for decades and as even secular geologists are starting to admit, we see much evidence for catastrophe in the rock record. From turbidites to tempestites to evidence of hurricane activity, etc., etc., *catastrophism is becoming the rule in geology.*

12

If indeed the rocks and fossils are the result of a global flood, then there is no evidence for evolution, nor for an old earth.

13

Fewer doctrines are taught with such clarity in Scripture as
recent creation and global flood.

one hundred one SIGNS of DESIGN

14

The fact that God created the various plant and animal types "after their kind," as repeated ten times in Genesis 1, absolutely prohibits them having descended with modification from previously existing kinds.

15

Truly, God's Word and God's world are both accurate self-authenticating and mutually reinforcing records of the unobserved past.

16

Obviously, fossils don't come with labels on them, explaining what each creature was like, where it came from, and when and how it lived. Fossils from the past must be interpreted.

17

History is reconstructed according to an individual's belief of the past, but this belief can never be proven; it is held for philosophical reasons.

one hundred one SIGNS of DESIGN

18

Life, from the simplest single-celled organism, to the largest dinosaur, to the human body, *exhibits a majestic design* that appears to be in direct opposition to random evolutionary change.

one hundred one SIGNS of DESIGN

The observed laws of science provide a huge barrier to evolution. A basic law of science, known as the "second law of thermodynamics," has been observed and verified in every field of science. This law reveals a general trend toward deterioration in all of nature. Everything moves in a downward spiral toward a less ordered state. Stars burn out. The moon's orbit is decaying. Cars wear out. Wood rots. Living plants wither and fade. Animals die. People get sick and grow old. No exception to this rule of decline has ever been observed.

20

Science operates in the present, and in a very real sense is limited to the present. Scientific theories must involve, among other things, the observation of data and process which exist in the present. But who has ever seen the long-ago past? Rocks and fossils exist in the present. We collect them, catalog them, study them, perform experiments on them — all in the present! The scientific method is an enterprise of the present.

21

Most people believe in evolution because most people believe in evolution.

one hundred one SIGNS of DESIGN

Contradictions in Order Between the
Biblical Order of Appearance

1. Matter created by God in the beginning
2. Earth before the sun and stars
3. Oceans before the land
4. Light before the sun
5. Atmosphere between two water layers
6. Land plants, first life forms created
7. Fruit trees before fish
8. Fish before insects
9. Land vegetation before sun
10. Marine mammals before land mammals
11. Birds before land reptiles
12. Man, the cause of death

one hundred one SIGNS *of* DESIGN

Biblical View and the Secular View

Evolutionary Order of Appearance

1. Matter existed in the beginning
2. Sun and stars before the earth
3. Land before the oceans
4. Sun, earth's first light
5. Atmosphere above a water layer
6. Marine organisms, first forms of life
7. Fish before fruit trees
8. Insects before fish
9. Sun before land plants
10. Land mammals before marine mammals
11. Reptiles before birds
12. Death, necessary antecedent of man

24

Ninety-five percent of all fossils are marine invertebrates, particularly shellfish. Of the remaining 5 percent, 95 percent are algae and plant fossils (4.75 percent). Ninety-five percent of the remaining 0.25 percent consists of the other invertebrates, including insects (0.2375 percent). The remaining 0.0125 percent includes all vertebrates, mostly fish. Ninety-five percent of the few land vertebrates consist of less than one bone. (For example, only about 1,200 dinosaur skeletons have been found.) Ninety-five percent of the mammal fossils were deposited during the Ice Age.

25

The stars were created for the purpose of being seen on earth, to accomplish the purpose of measuring time (Genesis 1:14-19).

one hundred one SIGNS of DESIGN

26

Suppose man has been around for one million years, as evolutionists teach. If present growth rates are typical there should be about 10^{8600} people alive today! That's 10 with 8600 zeros following it. This number is obviously absurd, and no evolutionist would claim it to be accurate.

There is plenty of water available to cover the earth. If the earth were completely smooth, the water would stand over a mile and a half deep.

Much of the Colorado River drainage basin is virtually untouched — a flat, featureless, uplifted plateau, bearing no evidence of 70 million years of erosion.

29

One animal *cannot become another.* God designed it that way. That's what the Bible means when it says God created all creatures *after their own kind.* When God created mice, their offspring years and years later were still mice. They didn't turn into human beings.

30

Every day, people discover gifts that God tucked into every corner of His creation. Gifts to make our lives easier or more enjoyable: from insects and sea animals, we learn how to produce inks and colored dyes; from birds, we design aircraft; from plants and animals, we make medicines; from bats, we learn radar.

God's gifts are endless. We need to remind ourselves that we never invent. We merely discover what God built or made possible when He created the world.

On tides: there are *two high tides each day,* when the moon is directly overhead or directly on the other side of the earth. The highest tide is called a *spring tide.* It happens when the moon and the sun are pulling in the same direction. The lowest tide is called a *neap tide.* It happens when the sun and the moon are pulling in different directions.

33

The largest animal is the blue whale. *The biggest one ever found was 100 feet long, weighing about 44,000 tons. This is bigger than the biggest dinosaur. The blue whale eats nothing but plants.*

34

The sperm whale has a huge head. The biggest ever found had a brain which weighed over 20 pounds. Although no one has ever seen him do it, he probably dives over two miles deep in the ocean.

The saltwater crocodile in southeast Asia and Australia can grow quite large and eat people! The largest killed was about 27 feet long and weighed over 4,000 pounds.

The heaviest fish:

an ocean sunfish once collided with a boat; the fish weighed almost 5,000 pounds.

The DNA in each cell tells the cell *how to grow and what to do.* It tells the cell whether to be a part of the brain or part of a finger. When a human baby is growing inside the mother's womb [the DNA] tells it to grow ten fingers and ten toes and one nose. It explains where the heart will be and shows the muscles how to make it beat. It knows if the child is to have red hair or blond, blue eyes or brown. The DNA knows it all.

38

Darwin, in his writings, letters, and memoirs, promoted natural selection as a means by which the incredible design obvious in every living system could be derived through purely mechanistic, naturalistic processes.

one hundred one SIGNS of DESIGN

If no supernatural agency has been at work throughout history, then creation is dead. But if evolutionists even allow a spark of supernatural design in history, then evolution is dead, for evolution necessarily relies on solely natural processes.

40

Design in living things is obvious.
Even the single-celled organism is complex beyond the ability of scientists to understand. The DNA code must not only be written correctly, the rest of the cell must be able to read it and follow its instructions, if the cell is to metabolize its food, carry out the myriad of enzyme reactions, and, especially, to reproduce.

41

A favorite example of obvious design has always been the human eye. With its many functioning parts — the lens, cornea, iris, etc., the controlling muscles, the sensitive rods and cones which translate light energy into chemical signals, the optic nerve which speeds these signals to a decoding center in the brain . . . the eye was unquestionably designed by an incredibly intelligent Designer who had a complete grasp of optical physics.

one hundred one SIGNS of DESIGN

42 *On the Yellowstone Petrified Forests:* Researchers have found an amazing diversity of plant species represented in the individual beds. Including pollen, up to 200 species have been identified from a wide range of ecological habitats, seemingly far too wide to have originated in one standing forest. Some species are from semi-arid deserts, others are from rain forests.

The study of tree rings provides insights into the history of a tree. Wet seasons, droughts, insect infestation, frost, and unusual weather patterns can all be discerned from tree rings. By comparing ring patterns from trees of overlapping life spans, a chronology of past events can sometimes be constructed.

While we can't be certain of the exact nature of the Flood, it certainly involved tsunamis — incredibly energetic shock waves in the ocean, traveling at the speed of sound, which pummeled the land with towering walls of water. Likewise, it involved underwater mudflows, which even today are known to flow at up to 100 miles per hour, following an underwater earthquake or other disturbance.

45

During the Flood, volcanism, tectonism, erosion, redeposition, etc., occurred at rates, scales, and intensities *far beyond similar processes occurring today.*

one hundred one SIGNS of DESIGN

46

Unfortunately, evolutionary illogic doesn't stop with biology, for "ideas have consequences." Justices see our Constitution as "evolving to fit the needs of a maturing society." Social engineers view man as an animal, complete with animal habits and tendencies. Politicians sanction promiscuity and homosexuality as animal behavior and as a beneficial mutation.

Since science and scientific method are limited to the present, how could fallible, limited scientists possibly reconstruct unobserved history? *Origins events are one-time, non-repeatable, unique events,* inaccessible to the scientific method. Empirical science, locked in the natural world as it is, can never succeed in reconstructing the supernatural acts of God without revelation, especially ultimate origins.

For years, evolution's twin theory was the principle of uniformity, that things have been "uniform" throughout the past. Advocated by Charles Lyell in the early 1800s, it became the "politically correct" position with its dictum "the present is the key to the past." Since the 1960s or so, strict uniformitarian thinking has relaxed somewhat, but still nothing too catastrophic is considered, such as the flood of Noah's day.

It has now been well demonstrated 49 *that rapidly moving, sediment-laden fluids can result in an abundance of laminations and/or layers. They can be formed in lab experiments, by hurricanes, and were even formed by catastrophic mud flows associated with the eruption of Mount St. Helens. A better interpretation of past deposits would stem from acceptance of rapid intense geologic processes, such as Noah's flood.*

50

Folklore has it, as reinforced in classrooms and national parks, that petrified wood takes "millions and millions" of years to form. I've listened as many people have protested the biblical doctrine of the young earth. "It takes too long to petrify wood. The earth must be old." Imagine their surprise when they realize that wood can petrify quickly, and that no informed geologist would say it takes an excessively long time, certainly less time than it takes for wood to decay in a given environment.

Wood can be petrified by two basic processes, both of which usually involve burial in volcanic ash. This ash decomposes in the presence of water, enriching the groundwater with silica.

A majority of today's evolutionists hold to the idea that a type of fish (order, Rhipidistia), led to amphibians. A "living fossil" fish, the coelacanthe, was found off the coast of Africa. This fossil fish had structures in its fins, and a loose comparison could be made with the femur and humerous (arm and leg bones in land animals), but nothing to compare to hands and feet.

53

The problem [of amphibians and land animals being linked] would be solved if we could find fossils of transitional forms, but alas, no "fishibian" has ever been found. Every fish, living or fossil, even those with unusual characteristics, is fully fish, and every amphibian, living or fossil, is fully amphibian.

54

A fossil amphibian has been found with "dates" even older than those "primitive" amphibians thought to be most fish-like. Yet it is 100 percent amphibian, just like it ought to be if (or should I say since) creation is true.

one hundred one SIGNS of DESIGN

A good definition of a miracle: an impossible event which happens anyway. Such a violation of natural law *requires supernatural input.*

Christianity demands miracles, and a miracle-working God. The miracle of creation, the incarnation, the Resurrection, the new birth. Without these miracles we could have no Christianity. With them we can have eternal life.

I suspect that *in His infinite wisdom,* God completely changed creation at the time of the Curse, with all things dying and some animals quite vicious, from then on giving eloquent testimony to the awful consequences of sin. From then on, whenever Adam saw one of the animals kill another, he would have experienced remorse for what he had brought on creation.

58

The Bible teaches that in the original creation, there was to be no death. Yet God also instructed Adam and Eve, as well as the animals, to be plant eaters. Would not the eating of plants constitute death? The answer to this seeming problem lies in a biblical understanding of "life," or "living." The Bible never refers to plants as living. They may "grow," or "flourish," but they do not "live." The Bible teaches that they may "wither," or "fade, but not die," since they are not "alive," having neither "life" (Hebrew nephish), nor breath of life (ruach).

In the modern land of Israel, many truly wonderful sights are on display. The coastal plain, the hill country, and the interior valleys all present striking, never-to-be-forgotten vistas. Geologic wonders are present in stark cliffs and folded strata, while fertile valleys and hillsides covered in springtime wild flowers provide arrays of color. But originally, God did not create the thorns and thistles, the dry wadies, the Dead Sea, and infertile rocky hillsides, the gnarled strata, the Red Sea-Jordan Valley transform fault, the abundant fossils, and the frequent lack of rain. These were the results of the Curse, a real, historical event.

Just imagine how beautiful God's original "very good" creation must have been, so that its destroyed remnant could still be beautiful.

The discovery of Noah's ark would be 61 the greatest archaeological find of all time. Some people think that it can be found or should be found. Others feel that it's not that important. Still others, myself included, think the search should continue. But the faith that God requires is not faith in the ark story or even blind faith in creation, the crucifixion, or the Resurrection. He has provided ample evidence for these great historic events, so that our faith is a reasonable one.

62

Someday, someone may possibly go into deep space, but the earth is our home. This is where we were born and where we live.

one hundred one SIGNS *of* DESIGN

As we observe this earth, we still see great beauty and marvelous balance, but we can never fully understand our present earth without knowing its past.

Only the earth is capable of supporting life.

65

If the earth's solid material were completely smooth, water would form a worldwide ocean approximately 8,500 feet deep!

one hundred one SIGNS of DESIGN

66

The interior of the earth is made up of four main sections. The crust is very thin and consists of the continents and oceans. The mantle is the largest at 1,900 miles thick. The outer core is so hot that it is molten liquid, while the inner core is under so much pressure that it is solid.

one hundred one SIGNS *of* DESIGN

The invisible magnetic field around the earth is a result of the earth having an iron core, in much the same way that an iron bar magnet produces a magnetic field.

The gravity of the moon

pulling on the earth causes the

oceans to rise and fall, forming tides.

69

The earth is the perfect distance from the sun to keep it the right temperature, and the earth's tilt causes the seasons.

one hundred one SIGNS of DESIGN

70

The earth's atmosphere not only provides air for breathing, but also deflects harmful space radiation and refracts solar radiation.

one hundred one SIGNS *of* DESIGN

Basalt makes up most of the oceanic crust, but can also be found on land.

Lava cools rapidly when it enters a body of water, forming "pillow lava," which resembles pillows stacked on top of one another.

73

Cave formations are primarily found in limestone. They were formed as groundwater evaporated, leaving its dissolved minerals behind.

one hundred one SIGNS of DESIGN

74

Chalk is made from the shells of single-celled organisms that have been deposited.

one hundred one SIGNS *of* DESIGN

Death Valley is a lake plain

that was formed as the water dried up,

leaving its sediments behind.

Rivers, *such as the Mississippi River,* carry great amounts of sediment in their waters. Sediments called alluvial sediments will be deposited whenever the water slows down, such as when it goes around a bend or as it enters the ocean, forming a delta.

77

When rock breaks and a large flat area is uplifted, then a fault plateau is formed.

one hundred one SIGNS of DESIGN

78

Areas such as Manhattan Island

were caused by a glacier scraping off

the sedimentary rock down to the granite core.

one hundred one SIGNS of DESIGN

Everything in the universe is subject to the law of disintegration. If processes continue as they do today, eventually everything will be eroded and worn away.

When water is flowing faster than 20 miles per hour, tiny vacuum bubbles are formed along the surface. These bubbles explode inwardly and pound the rock with great force, reducing it to powder in a process called cavitation.

81

The Tapeats Sandstone can be seen in the Grand Canyon as a dark flat layer. Scientists believe that it was formed by a series of underwater flows of sand. This thin "pancake" layer, the Tapeats Sandstone, covers most of North America, and would have taken a great underwater event in order to be formed.

82

The great majority of fossils are creatures with hard parts, like clams or coral, which lived on the ocean bottom. **These are preserved by the multiplied trillions.**

Trees can be turned into stone when they are buried in an area with silica-rich water.

It is hard to imagine, but there have been some fossilized examples of dung found in the rocks. The dung can be examined under a microscope and is found to contain remains of the organism's last meal.

85

Today, some birds eat tiny pebbles or sand grains to help them digest their food. In much the same way, dinosaurs or other large reptiles must have done this in the past, because on occasion a dinosaur fossil is found with many smoothly polished stones inside its body.

one hundred one SIGNS *of* DESIGN

86

Recent volcanoes, as destructive as they are, are not even comparable to volcanoes of the past. For instance, the eruption which produced Yellowstone National Park was tens of thousands of times more energetic than any we have witnessed in recent years.

one hundred one SIGNS of DESIGN

There is evidence that the continents were once connected into one large super continent called Pangea.

By summing up all the ways that salt enters and leaves the ocean, and measuring the amount of salt in the ocean now, scientists can approximate the maximum age of the ocean, and thus the earth.

89

Scientists can also date the earth

by measuring the total volume of the continents and how fast the sediments eroding from the continents enter the ocean.

one hundred one SIGNS *of* DESIGN

90

Before the Flood, it is believed that there was a dense cloud of water vapor that surrounded the earth and also a great deal of water underground. The dew watered the plants, and underground springs fed rivers and streams.

one hundred one SIGNS of DESIGN

From the flood account in Genesis, the "fountains of the great deep" might be best understood as great volcanoes and springs on the ocean floor.

The convulsions of the earth during the Flood would probably have caused the ocean basins themselves to rise, providing still more water which inundated the land.

We may see some faint reminders of these fountains of the great deep in modern-day springs in the ocean bottom. These bring up super-heated waters laden with a variety of chemicals, which are deposited on the sea floor as the hot waters encounter the cold ocean waters.

The earth suffered violent volcanic convulsions and rock movements so that major reforming of the continents took place during the Flood.

During the Ice Age, a glacier extended down into the present-day United States. Conditions were so harsh in part of the world that even the woolly mammoth could not survive.

As a glacier moves across the land, rock debris in the bottom of it scratches the bedrock below, leaving streaks called striations.

97

Marble Canyon and the Grand Canyon were most likely formed by a build-up of flood water in huge lakes that burst through the plateau, carving out the canyons.

98

Niagara Falls is the result of Lake Erie being higher than Lake Ontario and the draining water flowing over a rock cliff.

Dinosaur fossils such as the T-rex are actually quite rare.

100

In the original earth, the rains were restrained, but the abundant crops were watered from continual sources of mist and water springing up from below.

101

This present earth will be

completely melted and re-created (2 Pet. 3:10–13).

one hundred one SIGNS *of* DESIGN

Books by John D. Morris

Daddy Is There Really a God? Green Forest, AR: Master Books, 1997.

Dinosaurs, the Lost World, and You. Green Forest, AR: Master Books, 1999.

The Geology Book. Green Forest, AR: Master Books, 2000.

How Firm a Foundation in Scripture and Song. Green Forest, AR: Master Books, 1999.

The Modern Creation Trilogy (with Henry Morris). Green Forest, AR: Master Books, 1996.

Noah's Ark and the Ararat Adventure. Green Forest, AR: Master Books, 1988.

Noah's Ark, Noah's Flood. Green Forest, AR: Master Books, 1998.
Scopes: Creation on Trial (with R.M. Cornelius). Green Forest, AR: Master Books, 1999.
A Trip to the Ocean. Green Forest, AR: Master Books, 2000.
What Really Happened to the Dinosaurs? (with Ken Ham). Green Forest, AR: Master Books, 1990.
When Christians Roamed the Earth (with various other authors). Green Forest, AR: Master Books, 2001.
The Young Earth. Green Forest, AR: Master Books, 1994.

Available at Christian bookstores nationwide

For a free catalog or for more information about what the Bible teaches, contact one of the Answers in Genesis ministries below.

Answers in Genesis
P.O. Box 6330
Florence, KY 41022
USA

Answers in Genesis
P.O. Box 6302
Acacia Ridge DC
QLD 4110
Australia

Answers in Genesis
5-420 Erb St. West
Suite 213
Waterloo, Ontario
Canada N2L 6K6

Answers in Genesis
P.O. Box 39005
Howick, Auckland
New Zealand

Answers in Genesis
P.O. Box 5262
Leicester LE2 3XU
United Kingdom

Answers in Genesis
Attn: Nao Hanada
3317-23 Nagaoka, Ibaraki-machi
Higashi-ibaraki-gun, Ibaraki-ken
 311-3116
Japan

In addition, you may contact:
> Institute for Creation Research
> P.O. Box 2667
> El Cajon, CA 92021

one hundred one SIGNS *of* DESIGN

John D. Morris

Trained as a geologist (Ph.D. from the University of Oklahoma), Morris is one of the world's most visible authorities on creation science. As president of the renowned Institute for Creation Research in San Diego, Morris writes and speaks extensively. He and his wife, Dalta, have four children.

one hundred one SIGNS *of* DESIGN